A COMMENTARY ON THE MUTUS LIBER

MUTUS LIBER, IN QUO TAMEN
tota Philosophia hermetica, figuris hieroglyphicis
depingitur, ter optimo maximo Deo misericordi
consecratus, solisque filiis artis dedicatus,
authore cuius nomen est Altus.
21. ii. 82. Neg:
93. 82. 72. Neg:
82. 81. 33. Tued.

A COMMENTARY
ON
THE MUTUS LIBER

By Adam McLean

MAGNUM OPUS HERMETIC SOURCEWORKS #11

PHANES PRESS
1991

My grateful thanks to Christopher McIntosh for providing a translation of 'The License of the King' appended to the early editions of the Mutus Liber.

This work, part of the Magnum Opus Hermetic Sourceworks series, was previously published in a handbound edition, limited to 250 copies, in 1982. The Magnum Opus Hermetic Sourceworks series is published under the General Editorship of Adam McLean.

98 97 96 95 94 93 92 91   5 4 3 2 1

Published by Phanes Press, PO Box 6114, Grand Rapids, MI 49516, USA.

Library of Congress Cataloging-in-Publication Data

McLean, Adam.
    A commentary on the Mutus Liber / by Adam McLean.
        p.   cm.
    ISBN 0-933999-89-5 (alk. paper) — ISBN 0-933999-90-9
(pbk. : alk. paper)
    1. Mutus liber. 2. Alchemy—Early works to 1800. I. Mutus
liber. 1991. II. Title. III. Series: Magnum Opus Hermetic
Sourceworks (Series) ; no. 11.
QD25.M37   1991
540'.1'12—dc20   90-47419
                                                   CIP

This book is printed on alkaline paper which conforms to the permanent paper standard developed by the National Information Standards Organization.

Printed and bound in the United States

Contents

Introduction 7

The Fifteen Plates of the Mutus Liber with
   Descriptive Summaries 11

Commentary 47

Introduction

The *Mutus Liber* is one of the most reproduced series of alchemical illustrations; however, it remains essentially mysterious and enigmatic. It was first printed in 1677 at the instigation of a Frenchman, Jacob Saulat, though he was not the author of the work. Authorship is assigned to an anonymous figure named 'Altus'— 'the high, deep or profound one.' In a document called *License of the King*, which was attached to the first printing and follows this introduction, it is stated that "there had fallen into his [Saulat's] hands a book of high Hermetic Chemistry entitled *Mutus Liber*." This suggests that he had somehow acquired a manuscript copy of the work, recognized its essential value, and sought to make it available to the public. It is interesting that he felt he needed the copyright protection of the King and his Court, and this indicates that there must have been considerable interest in such alchemical material in France at that time, enough that he was worried about some printers producing pirate editions. Perhaps he paid a considerable sum for the original manuscript. This first edition is exceedingly rare, though I have been able to examine a copy in the Ferguson Collection at Glasgow University Library.

The *Mutus Liber* was, however, made widely known through its inclusion at the end of the first volume of Manget's MANGETUS (Joannes Jacobus, editor, *Bibliotheca Chemica Curiosa, seu Rerum ad alchemiam pertinetium thesaurus instructissimus*, Geneva, 1702, 2 volumes). This alchemical compendium was widely available and helped to establish the reputation of many alchemical works, including the *Mutus Liber*.

The plates in Manget's edition are of superior execution, being more finely engraved than Saulat's edition, and it is this set that has been reproduced here.

There are no great variations between these two editions, merely the superior artistic merit of the Manget engravings, and the fact that Saulat has 13 plates instead of the 15 found in Manget.

Up until plate 11 both series correspond exactly, then Manget has two extra plates numbered 12 and 13 (almost exactly mirroring

9 and 10). His series concludes with plates 14 and 15 which correspond to Saulat's numbers 12 and 13. There are two obvious interpretations of this fact. Either Saulat's edition was the original source for Manget, in which case Manget included two extra plates of his own invention, or Manget had access to another original manuscript that included these two illustrations. As these two plates, Manget 12 and 13, are very similar to Manget and Saulat 9 and 10, it could also be that the printer of the Saulat edition was deceived, thinking them to be alternative versions of the same plates, and thus felt them to be redundant. However, as I indicate in my commentary, though both editions admit of coherent symbolic interpretations, Manget's extra plates indicate a repeated cycling of the process; thus I have chosen to take Manget's edition as the complete version of the work. To my knowledge, no early manuscripts of the *Mutus Liber* that one could compare with these printed versions are extant.

Whatever its early history, the *Mutus Liber* made a considerable impact upon later alchemical studies, particularly in the French occult schools. Perhaps the seeming simplicity of its symbolic statement, or its detailed description of the physical alchemical process, attracted the attention of later commentators and students of alchemy. It seems almost like an alchemical 'strip cartoon,' with the step by step stages in the work pictured in a definite sequence.

A facsimile edition of Manget's engravings was published in France during the early decades of this century, and this sparked off an interest in the work among twentieth century students of alchemy. Fulcanelli mentions the work in his *Les Demeures Philosophales*, and his pupil Canseliet published a commentary in 1967 on the *Mutus Liber* entitled *Mutus Liber L'Alchimie et son livre muet, Mutus Liber: Reimpression premiere et integrale de l'edition originale de La Rochelle 1677*.

Perhaps the most important of modern alchemists to have been inspired by the *Mutus Liber* was Armand Barbault, whose *Gold of a Thousand Mornings* was first published in France in 1969, and in an English translation by Robin Campbell, published by Neville

Spearman in 1975.

Barbault gained inspiration by meditating on the figures of the *Mutus Liber*, and though he did not attempt to reproduce the process exactly as outlined, he took certain facets of the *Mutus Liber* process as starting points for his own experiments. For example, he collected dew using a method similar to that outlined in plate 4, using canvas cloths from which the dew could be wrung. This dew, which contained the etheric essence of the forces of spring, was subjected to a subtle process of purification and used to bathe or nourish his primal material. For this primal matter Barbault used a sample of raw earth, to which he was guided by the inspiration of his wife. They dug up this raw earth, rich in etheric forces, under her guidance and during certain favorable astrological aspects calculated by Armand Barbault. During many seasons of washing and purifying this earth, 'the Philosophical Peat,' it was transformed into a golden elixir, 'The Gold of a Thousand Mornings.' An important aspect of the whole process was working in partnership with his wife. They shared the work and each contributed special gifts and abilities as man and woman to the task. This mirrors the cooperation of the male and female in the series of the *Mutus Liber*. The Barbaults' work shows that there is, indeed, still life in this series of engravings and that it continues to inspire and illumine those who work with its essence in an alchemical operation.

I have found the Barbaults' work to be of great help to me in grasping something of the inner content of the *Mutus Liber*, and I hereby acknowledge the inspiration his book has given to many points in my commentary. I would recommend *The Gold of a Thousand Mornings* as essential parallel reading material to those attempting to understand the *Mutus Liber* and urge them to work with it both symbolically and practically.

Another Frenchman, Jean Laplace, produced an edition of the *Mutus Liber* entitled *Altus, Mutus Liber, Reproduction des 15 planches en couleur d'un Manuscrit du XVIllème Siècle*, Introduction et Commentaire par Jean Laplace, Milano, Arche, 1979. It was a color reproduction of a French manuscript, the original of which was donated to the Library of Congress by Mrs. H. Carrington Bolton of New York in June, 1914 (MS. Div. 1507). Laplace indicates that the original may date to 1760. This colored manuscript is of great beauty and interest, though it seems that the colors are not symbolically important and that the actual drawings are what give it significance. Therefore I feel quite sure that the essence of the work is contained in the drawings alone which can quite readily be found in the engraved version of Manget.

It is interesting that the *Mutus Liber* seems not to incorporate the usual sequence of Nigredo–Albedo–Rubedo, i.e., the Black, White and Red stages of the work, but approaches the alchemical process from another angle, and grasping the essence of this *Mutus Liber* process seems to necessitate an etheric viewpoint. Thus the *Mutus Liber* presents us with a process revealing alchemy as an etheric science.

For my commentary, I have chosen to reprint the fifteen plates from Manget's edition, with a summary of the main symbols on each plate on the facing page. The interpretation of these symbols in sequence constitutes the main body of the commentary. As with similar commentaries I have provided in other volumes of the Magnum Opus series, it is not my intention that my readers see this as a 'final answer' to the riddles of the work, but rather use it as a tool for working with the *Mutus Liber* process itself. Only through active work and a personal encounter with the *Mutus Liber* will inner understanding of its essence be won by the aspiring student. Since different students from various backgrounds, each with their unique abilities, may discern different facets of its essence, there will be many possible valid interpretations. This multiplicity of interpretative potential is a consequence of the *Mutus Liber*'s stature as a major alchemical work. I hope my commentary will help people to explore these many dimensions of the process, adding something to the *Mutus Liber*, and not stripping from it any of its mystery.

—ADAM MCLEAN

THE FIFTEEN PLATES
OF THE
MUTUS LIBER

with
Descriptive Summaries

License of the King

Louis, by the Grace of God, King of France and of Navarre: To our beloved and faithful Councillors, Gentlemen of our Courts of Parliament, Masters of ordinary requests of our household, Grand Council, Provost of Paris, Bailiffs, Seneschals, Provosts, their Lieutenants and other Judges whom it will concern; GREETINGS. Our well-beloved Jacob Saulat, Seigneur of Marez, has made it known to us that there has fallen into his hands a book of high Hermetic Chemistry entitled: *Mutus Liber, in quo tamen tota Philosophia Hermetica Figuris hieroglyphicis depingitur, ter optimo, maximo Deo misericordi consecratus, solisque Filiis Artis dedicatus, Authore cujus nomen est Altus*; which he desires to present to the Public. But he fears that, after the expense has been met by himself or some bookseller or printer of his choice, others will undertake to print it unless he has our necessary letters. FOR THESE REASONS, wishing to oblige the applicant, we have permitted and granted, and do permit and grant by the present document, that the book be printed in such manner as will seem good to him, and that within the time and space of ten consecutive years, to begin on the day when the first printing is finished, it be sold and supplied throughout our Kingdom. We forbid all booksellers, printers and others, to print, have printed, sell or supply the said book under the pretext of augmentation, correction, change of title, or foreign edition, or in any other way or manner, and even to make an extract from it, without the permission of the applicant or those acting for him, on pain of the confiscation of these counterfeit copies, a fine of two thousand pounds and the payment of all expenses and damages. In the case of contravention we and our Council shall take cognizance of the fact. Two copies must be placed in our public Library, one in the Book Cabinet of our Palace of the Louvre, and one with our very dear and faithful Seigneur d'Aligre, Chancellor and Guardian of the Seals of France, and this document must be registered with the Community of Paris booksellers, otherwise it will become null and void: We order and enjoin you to give the benefit of the contents of this document to the applicant and those acting for him, fully and peacefully, ceasing and causing to cease any troubles and impediments to the contrary. We desire that, in placing an extract from this document at the beginning or end of the said books, they shall be deemed to be legally signified; and that as much credence be given to the copies verified by our beloved and faithful Councillors and Secretaries as to the Original. We command that our Bailiff or Sergeant sees to all the necessary notices, prohibitions, seizures and other actions necessary for the carrying out of this order, notwithstanding the Clameur (Proclamation) of Haro, the Norman Charter or other Letters to the contrary, for such is our pleasure. Given at Saint Germain on the 23rd day of November in the year of grace 1677, and the 34th of our reign. Signed by the King in his Council: DESVIEUX.

The said Seigneur Saulat has permitted Pierre Savouret, bookseller at La Rochelle, to print, sell and supply the said book, according to the agreement made between them.

First Printing completed on 1st February 1677.

Registered in the Book of the Community of Booksellers and Printers of Paris, on 28th November 1676, following the judgement of Parliament of the 6th April 1653 and that of the King's Privy Council of 27th February 1665. THIERRY, Syndic.

The copies have been supplied.

THE PLATES

Plate One

The opening illustration to the *Mutus Liber* shows a Jacob's Ladder contained within a piscina or oval formed by two rose branches. It is set up against the dark heavens in which the stars and moon shine out, reaching down to the Earth below. Two winged angels ascend and descend this ladder, blowing trumpets to arouse the sleeping figure at the bottom right of the illustration.

The short title reads: "The wordless book, in which nevertheless the whole of Hermetic Philosophy is set forth in hieroglyphic figures, sacred to God the merciful, thrice best and greatest, and dedicated to the sons of the art only, the name of the author being Altus."

The series of numbers and letters at the bottom are actually biblical references in reverse. Thus:

21	11	82	Neg	is Genesis 28:11, 12.
93	82	72	Neg	is Genesis 27:28, 39.
82	81	33	Tued	is Deuteronomy 33:18, 28.

And he came to a certain place, and stayed there that night, because the sun had set. Taking one of the stones of the place, he put it under his head and lay down in that place to sleep. (Genesis 28:11)

And he dreamed that there was a ladder set up on the earth, and the top of it reached to heaven; and behold, the angels of God were ascending on it! (Genesis 28:12)

May God give you of the dew of heaven, and of the fatness of the earth, and plenty of grain and wine. (Genesis 27:28)

Then Isaac his father answered him: "Behold, away from the fatness of the earth shall your dwelling be, and away from the dew of heaven on high. . . ." (Genesis 27:39)

And of Zebulun he said, "Rejoice, Zebulun, in your going out; and Issachar, in your tents. . . ." (Deuteronomy 33:18)

So Israel dwelt in safety, the fountain of Jacob alone, in a land of grain and wine; yea, his heavens drop down dew. (Deuteronomy 33:28)

MUTUS LIBER, IN QUO TAMEN
tota Philosophia herme- tica, figuris hieroglyphicis
depingitur. ter optimo maximo Deo misericordi
consecratus, solisque filiis artis dedicatus,
authore cuius nomen est Altus.
21. ii. 82. Neg:
93. 82. 72. Neg:
82. 81. 33. Tued.

Plate Two

This divides into two sections.

In the upper section, two winged angels, whose feet touch upon a watery sea, hold aloft a sealed flask. Within it is pictured Neptune with his trident, seated upon a rock at the foot of which a dolphin is seen. Beside him, standing under the protection of his outspread arms are two clothed children bearing Sun and Moon symbols.

In the lower part of this plate, a couple of alchemists kneel in front of their furnace. Heavy curtains obscure the background of their work. The furnace is designed like a castellated tower with three interior parts. At the base a lamp burns to provide a slow steady heat; in the middle of the tower furnace a funnel-shaped device is seen; and immediately above this is a hermetically sealed flask. The male alchemist on the left kneels in passive prayer, while his companion, the female alchemist on the right, is more animated, as if trying to communicate some insight or inspiration; her prayer for the success of the work is a more active exhortation.

19

Plate Three

In the spiritual realm far above the starry heavens of Sun and Moon, we see Zeus-Jupiter seated upon his Eagle, gazing down upon the Created World. This globe is divided into three concentric circles or three realms. The outer circle shows Juno, Jupiter's sister, wife, and female syzygy, with her Peacock symbol (mirroring his Eagle). She points towards a group of ten birds that fly in the airy region, mediating between the upper spiritual sphere and the Earth. The surface of the Earth itself is seen in the central circle, where a female Earth Mother figure, seated upon a mound, tends a plant with five blossoms in a pot. Beside her, upon the mound, a plant with five blossoms grows directly in the natural environment. A Ram and a Bull occupy opposite sides of this circle, the Ram on the Solar side, the Bull on the Lunar. Also seen associated with the Solar side is a castellated tower, symbolic of the outward power of the masculine might in human society, while on the Lunar, receptive side the Church is pictured, the power of organized religion which rules humanity inwardly and requires the imaginative, devotional side of our human nature for its existence.

In the lower part of this middle circle of the Earth sphere, two figures, a female on the left with a net and birdcage and a male with a fishing rod, attempt to catch the mermaid creature in the sea at the bottom of the outer circle. In the central circle is also pictured a sea realm within which a couple in a boat, the male rowing and the female with a rod, attempt to catch something out of the sea. Within this is seen Neptune in a sea-chariot drawn by sea-horses, and he has a line in his hand connecting him with the woman in the boat.

The earth domain lies between two spheres: an upper spiritual realm in the heights above—the Zeus kingdom, mediated through the Air element by the alchemical birds—and a lower, dark watery depth, within which lurk spiritual forces personified in Neptune and his mermaid spirits. The alchemist must net, catch and encounter both the spiritual forces from the watery depths below the Earth and those from the airy heights above. This macrocosmic picture also applies to the microcosmic world of man, and to his experiments in the sealed flasks.

Plate Four

The physical task of the alchemist in the process begins here with the collection of dew, which is accomplished by stretching large square cloths across four wooden posts, thus keeping them taut and horizontal just above the ground surface. The dew forms upon these cloths and saturates them, and can be gathered by wringing them out. The two figures of the Alchemist and his Soror Mystica are seen doing this at the bottom of the plate. Six cloths in all are shown here, which must have been arranged in a triangular array ∴. . They have been placed in a meadow and we see the Ram and the Bull on opposite sides of the plate. This could indicate that the collection of dew should occur during the spring signs of Aries and Taurus (April-May).

Above in the heavens we see the Sun and Moon, and from the zenith, the height of the heavens, streams of forces descend to the Earth, radiating downwards. This is the tide of etheric forces which begins to descend to Earth during the springtime. It is this cosmic tide that unites with the rising dew from the Earth and charges it with etheric force. Since it is these incarnated etheric forces that the alchemists wish to work with, they must collect the dew.

23

Plate Five

Here we see the alchemical couple beginning their work on the prima materia, which is the dew collected in the previous stage. This dew, wrung out of the cloths onto a large flat plate, is in turn placed in a vessel set into a furnace. Both partners work together in this action. In the next picture the female partner fits a still head to the vessel, while the male partner holds a receiving flask to collect the distillate. A fire is lit in the furnace and the dew is distilled. In the third picture in this strip, the male partner, still holding the collecting flask, removes the still head of the alembic, and allows his female helpmate to remove the concentrated residues in the vessel with a spoon. This residue, which is symbolised by △◠△, she places in a bottle. In the next picture this square-bottomed bottle is handed to a Kronos-Saturn figure standing upon a slight mound and holding a child. (Kronos-Chronos is often pictured in alchemical illustrations in the act of devouring his children.) This figure bears the lunar crescent which is also seen engraved upon a shield beside him.

With the next picture on this plate a new stage opens in which the male alchemist places the distillate, collected in his flask during the previous process, into small circulating vessels for reflux. These are heated by a furnace which does not provide a fierce direct heat, and probably a water bath or Balneum Maria is intended here. The female alchemist fits the heads onto the four cucurbites or reflux vessels. At the bottom, the number 40 indicates that this process should be continued for forty days.

25

Plate Six

The first picture portrays how the process outlined in the previous plate continues once the fire is extinguished. The male alchemist pours the contents of the four circulating vessels into a larger flask which the woman alchemist holds for him. This is then placed in a furnace upon a water bath and fitted with a still head and collecting vessel. In the third image, we see the furnace lit and the water bath steaming and heating the alembic. The woman holds the still head while the man secures a receiving flask to collect the distillate. Within the distilling flask a flower slowly forms. In the fourth figure, the man removes the still head and allows the woman alchemist to collect the residue with a spoon before placing it in a round-bottomed flask. This residue bears the symbol ✿ of a flower. The fifth figure shows the male alchemist handing this flask containing ✿ to an Apollo-Sun figure.

The next stage begins with the last image on this plate: here the female alchemist places the residue from the first experiment, the ⌒⌒ contained in the square-bottomed bottle, into a crucible heated on the furnace, in order to evaporate it to dryness.

27

Plate Seven

The first image continues the process initiated in the last image of plate 6. Here the ⌂⌂⌂⌂ residue, heated to dryness, is placed in a shallow vessel and wetted with one of the flasks of distillate held by the male alchemist, while the woman pounds and grinds the mixture with a pestle. In the second picture, this homogenized mixture is poured through a funnel into a large flask; later, in the third image, this is evaporated slowly to dryness in an open shallow vessel held over a water bath set upon the furnace. The male alchemist pours the liquid into this vessel, while his female partner holds the funnel aloft. In the fourth figure, the furnace being extinguished, the woman gathers the residue with a spoon and places it in a square-bottomed flask. This residue here bears the symbol ⁂⁂.

The concluding series of three images in the lower part of this plate recapitulates in an allegorical manner the three stages outlined in plates 5, 6 and 7. Firstly, Kronos-Saturn, seen in the act of devouring his child, is subjected to a fiery purification; next he is pictured in his tub being washed in the liquid distillate. Finally he cuts a cord binding him to the bottle with ⁂⁂, thus freeing it from himself and letting it pass into the hands of Luna ☽, the Moon Goddess.

Plate Eight

Although this plate is similar in form to plate 2, certain important differences can be noticed. In the upper part of the plate, two angels hold aloft a sealed flask, which on this occasion bears within it the figure of Mercurius holding a strange caduceus with eight snake heads. At his feet within an earthy deposit are seen the symbols of the Sun and Moon archetypes. This whole event takes place not upon the water, but entirely in the air. The ten birds of plate 3, here in two flocks of five on either side of the flask, each have a leader who bears a branch in its beak. These branches have symbols attached to them. That on the left bears ⚷ and that on the right ✳. The Sun above presides over these events.

Below, in the alchemist's chamber, our male and female alchemists again kneel in front of their furnace, the vessel of transformation. At this stage no lamp is lit at the base of the furnace, though the flask is seen intact at the top. The male alchemist again is seen in an attitude of passive prayer, while his female counterpart seems to be more active, perhaps speaking out of inner inspiration. Behind them, the heavy drapes have been somewhat parted to reveal two rounded windows from which they can look out from their chamber. Also two pillars have appeared on either side, uniting the above and the below.

Plate Nine

Here we return to the landscape of plate 4. The Sun and Moon gaze down on a meadow, at opposite ends of which the Ram and the Bull are seemingly set to battle. Upon this meadow are set six shallow open vessels, arrayed in a triangular form.

Below, the female alchemist is seen pouring the contents of one of these plates into a receiver held by her male companion. She then hands this flask of liquid to Mercurius.

33

Plate Ten

The first picture in this plate shows the male alchemist pouring equal parts of the ⁂ from a square-bottomed flask held in his right hand, and ⚛ from the round-bottomed flask in his left hand, into the two pans of a balance, so that they are of exactly equal weight. His female counterpart pours these contents of the balance pans into a round-bottomed flask. The next picture shows the man pouring the contents of another flask (probably the substance collected in plate 9) into the flask with ⚹ and ⚛ held by the woman, and in the following figure he hermetically seals the neck of this flask by heating it in a flame. He uses a blowing tube to create a higher temperature flame in order to melt the glass more easily, for once the glass neck of the flask is soft it can be twisted and sealed. The fourth picture on this plate shows the placing of this flask into the furnace of transformation, which corresponds in form to the furnaces shown in plates 2, 8 and 11.

The last image is an allegory of the conjunction of Luna and Sol, the Moon and the Sun. Each have beside them the number 10, and beside the furnace, which has a lamp burning at the base to provide a gentle heat, is a cross section of an egg with four layers.

Plate Eleven

Plate 11 corresponds in form to plates 2 and 8, though we note some important differences.

The upper part of the illustration shows two angels bearing aloft the flask containing Mercurius with his strange caduceus and winged helmet. In distinction from plate 8, the earthly substance at the base of this flask has disappeared, being dissolved into Mercurius, though he still stands upon the Sun and Moon symbols. The birds again bear gifts. In this instance, the bird on the left has as an addition to ⍦, the symbol ☋ (sublimated ⍦, ⍦) while the bird on the right still bears ✳.

Below, in the alchemists' chamber, the couple still kneels in front of their furnace of transformation, which has a lamp burning at its base, and the female again displays the more active gesture. The heavy drapes in the background have been removed and they can now look out through two additional oval windows.

Plate Twelve

This corresponds almost exactly to plate 9. Again six shallow vessels are placed in a triangular array upon a meadow where the Ram and the Bull contest. Beneath this scene the female alchemist pours a liquid from these vessels into a flask held by her male partner, and later she hands this to Mercurius.

39

Plate Thirteen

This pictures a similar conjunction to that depicted in plate 10. However, this time the male figure pours into the balance pans equal parts of ⚹ and ☀, which are then amalgamated in the flask by his female counterpart. Next, he adds a liquid from a round-bottomed flask. This is probably what was collected in plate 12. This flask is again sealed using a blow torch flame and placed in the furnace of transformation.

Below, in the last figure, the Conjunction of Moon and Sun is shown. This time the figures "100 . 1000 . 10000 etc." appear beside them. Again the diagrammatic cross section of an egg is shown, this time somewhat larger than in plate 10.

Plate 13 thus shows another conjunction of Sun and Moon in the furnace of transformation, which again has a lamp under it providing a gentle heat.

Plate Fourteen

At the top we see the three furnaces of transformation with their lamps burning below. Behind, the heavy drapes we noted before in plates 2 and 8 are drawn aside and two windows looking out from this chamber are seen.

In the next picture below, three figures—the Man (with torch), the Child (shown with tennis racket and ball), and the Woman (also with a torch)—are at work tending their lamps, trimming the wicks and filling them with oil. Beside them are the roman numerals VI (next to the man), II (next to the child), and X (next to the woman), possibly indicating the number of months that the process must be continued.

In the third figure, two square roaring furnaces heat tightly sealed crucibles to a high temperature. The one on the left contains the Lunar Tincture, the one on the right the Solar Tincture. In the center, a balance indicates that equal parts of each of these should be taken and ground together in a mortar which is engraved with a shell and which has two snakes at its base; to its left and right are seen the balance weights.

In the final figure below, the result of this work is shown in the flask bearing the Philosophers' Mercury. The male and female alchemists gesture upwards but seal their lips. Between them, below the furnace tongs, we see the exhortation—"Pray, Read, Read, Read, Read again, Labor and Discover."

Ora
Lege Lege Lege Relege labora
et Invenies.

Plate Fifteen

The final Plate in this series is a completion of plate 1. Again two branches form themselves into an oval, though this time they are not thorny rose bushes but branches bearing a triple fruit. Jacob's Ladder is no longer necessary and lies horizontal in the background. The two alchemists have achieved spiritual illumination and can ascend to the spiritual realm without the angels of Jacob's Ladder rousing them to insight. The male figure and the female figure reach out and touch right and left hands respectively, while they grasp with their other hands two cords let down from the Zeus figure who is crowned with a laurel wreath by the two winged cherubim that appeared in plates 8 and 11. From the mouths of the alchemists two banners proclaim "provided with eyes thou departest." From the two cherubim the red and the white rose emanate toward the man and the woman.

Between the man and woman are the archetypal symbols of Moon and Sun, while below them is a Hercules figure clad in a lion's skin and bearing a club. This may refer to the myth of the Nessus shirt that caused the death of Hercules.

At the bottom of this illustration is seen a shield within an oval of leaves, bearing important symbols of the alchemical process.

Commentary

The *Mutus Liber* is an enigmatic work, as befits one of the most significant statements of the alchemical process. The important documents of alchemy, the texts and series of symbolic pictures, become enigmatic as they come close to revealing the nature of their subject, dissolving the clear statement of ideas into unfocussed obscurity and paradox. In a textual work, this often finds expression in alchemical allegory, where the text obliquely touches upon the subject, pursuing it through an elaborate and often tortuous allegory, tantalizing the reader with glimpses of clear statement. In a series of alchemical illustrations, the enigma is woven into pictorial symbols which seem to relate together easily, but on deeper consideration somehow possess a dynamic symbolic energy in their relationship, giving these pictures an inner tension that disturbs the soul. It is this enigmatic quality that gives alchemical material an on-going life.

If indeed the message of alchemical documents was revealed straightforwardly, requiring no inner effort to unravel the mystery, our interest and the on-going value of these texts would rapidly wane and they would become like dead corpses. The ancient alchemists fully realized this fact, and wove subtle mysteries, through allegory, paradox and symbolic tension, that would project into the soul of those who later studied these texts and awaken there a sense of wonder, an inner need to pursue the enigma of the alchemical process they sought to preserve.

The *Mutus Liber* is such an alchemical document. It is entirely symbolic, consisting of fifteen engraved plates, and presents its mystery through a seeming simplicity of statement. On first examining this work, it appears to show quite clearly, step by step (like a strip cartoon), just how to proceed in the alchemical process. However, when we work deeper and deeper into the symbolism we will find that seeming simplicity in fact hides a multiplicity of levels of interpretation. Many who have tried to unravel its mystery find, even within its small compass, that its essence seems to dissolve before their intellectual grasp into manifold facets.

This work was originally published in 1677 without any commentary and has been reprinted many times since. However, I feel that it now requires a commentary that can help to unravel some of these manifold facets. My method will be to point out and amplify some of these facets—these esoteric threads running through its symbolic fabric—without attempting an exhaustive statement on the work. It will not be my task here to explain the work or attempt to reduce it to a simplistic statement, but rather to point out ways of approaching its mystery and for working further with its ideas.

It is esoterically wrong to attempt to strip a work of its mystery, its inner life, and I will not pursue this course here, even if I were capable of it. Certainly, in the case of the *Mutus Liber*, I can find no single simplistic statement of its message, so such a reduction-ism is here for me ruled out, even if I wished it. So I trust that this commentary will help to reveal the enigmas of the work, and allow the reader to further ponder on its mystery, without losing that subtle aura of wonder that exists around this precious document of alchemy. I find a kind of life still active within this set of fifteen illustrations that can quite easily be committed in sequence to memory, becoming an inward possession, an inner book, to medi-tate upon occasionally. Brevity and conciseness of expression is perhaps one of the *Mutus Liber*'s most important facets, and unlike many alchemical works its symbols can easily be inwardly re-called.

Let us now look at the underlying structure of the work. In other commentaries on series of alchemical pictures, I have ini-tially sought a pattern or arrangement of the pictures, breaking them down by this analysis into simpler units, and then reintegrating these into a whole, through seeking a synthesis in interrelationships. In the case of the *Mutus Liber* series, this task is made easy as a number of the illustrations are very similar, though not identical, in form (for example plates 2, 8 and 11 ; and plates 4, 9, and 12). This immediately gives us a key to the underlying structure:

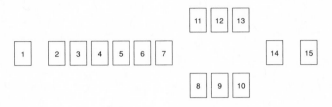

We have an initial process outlined by plates 2-7, then two Conjunctions pictured by 8-10 and 11-13, which are on different levels, and a final synthesis in plate 14 completes the work.

This alchemical process is sandwiched between the two outer plates 1 and 15, the beginning and the end of the work, and these indeed provide us with a key to the inner content of the process. So we will look initially at these two plates.

The Beginning and End of the Work

Plates 1 and 15.

These two illustrations, within which the rest of the figures are, as it were, enfolded, ensure that we note the spiritual parameter of the alchemical work. Some other plates seem to be descriptive of physical processes only, and so the creator of this series has ensured that the spiritual aspect is not forgotten, by giving both the beginning of the work, where the intention is revealed, and the end of the work, where the outcome is pictured, a spiritual interpretation.

These two plates correspond closely in form. The outer part of both figures is an oval (or piscina) formed by two branches tied together at the base. In plate 1, this is formed by two thorny rose branches, while in the last plate this is formed by two fruit-bearing branches.

Plate 1 is a cry to awaken to awareness of the potentialities of the work. A sleeping figure rests with his head upon a rock, and beside him a Jacob's Ladder is set up bridging heaven and earth. Upon this ladder two angels, spiritual intelligences, seek to awaken the sleeper to a consciousness of the spiritual by blowing their trumpets. The first biblical quotation on plate 1 focusses our attention to this point.

> And he came to a certain place, and stayed there that night, because the sun had set. Taking one of the stones of the place, he put it under his head and lay down in that place to sleep. (Genesis 28:11)
>
> And he dreamed that there was a ladder set up on the earth, and the top of it reached to heaven; and behold, the angels of God were ascending on it! (Genesis 28:12)

It is only by an awakening to the spiritual potentialities of matter that the alchemist will be able to embark upon his task with any hope of success. We note also that these events take place at night, the crescent moon shining far above, with the stars also shown against the darkness of the sky.

The rose branches illustrate that the beauty of form found in the rose flower arises from an unpromising thorny bush. The rose unites beauty with its harsh thorns. The alchemist must often work with hard, unpromising material, and seek there the spiritual potential, the inner rose.

Having seen that the intention of the work is to seek the spiritual potential in the material, and that this requires an awakening to the lunar and stellar forces of the dark night, we now move to a consideration of the final plate, the outcome of the work.

In contrast to plate 1, the happenings here take place in bright sunlight, the Sun shining at the top center of the piscina. This oval is formed by two branches without thorns which bear round fruits. Thus we have moved from the flowering stage of spiritual awakening to the fruition of the work. In this last plate, the Jacob's Ladder is no longer needed and lies horizontal, unused in the background.

Throughout the *Mutus Liber* a pair of alchemists, male and female, are shown jointly participating in the work, and here the alchemical couple is ascending to a spiritual state, their inner eyes having been opened (the banners state "provided with eyes thou departest"). In distinction to the angels of plate 1, here it is Zeus that lets down a cord to raise the alchemical couple. (Note that it would be entirely erroneous to identify this being as a Christ figure, as we see him depicted clearly as Zeus in plate 3, and the mythological background to the *Mutus Liber* is that of classical mythology, and it does not seem to incorporate any Christian elements in its symbolism).

This event is pictured with an underlying square form as the alchemical couple reach out, and by holding each other's hand complete the side of the square formed by the cord. The male alchemist reaches out his right (active-dexterous-male) hand to grasp his partner's left (passive-sinister-female) hand. The male alchemist reaches up to the cord of the spiritual with his left (passive-sinister-female) hand and his female partner grasps the spiritual cord with her right (active-dexterous-male) hand. Thus the spiritual side of our nature can be found by exploring the opposite side of our psyche than that which is incarnated in our physical sex. The active, dexterous side of the male tends to touch the outer world, and he who narrows his psyche to this facet of his being can only grasp the material side of life. He must awaken to his inner, feminine side if he is to reach out to the spiritual. Similarly, in the case of the female, she must seek her active side in order to realize her relationship to the spiritual. The mutual grasping of each other's hand remains within the human sphere, as the male puts out his right and the female her left in their earthly handfasting.

The Zeus figure is crowned with a laurel wreath by two cherubim, and together they form another square arrangement in the higher spiritual plane. The two roses, seen on the first plate, are now pictured emanating from these cherubim, indicating that the impulse behind the initial flowering of consciousness had its essence in the spiritual, as in the end of the work.

Below the two alchemists, we see the reason for their elevation to spiritual awareness in the Moon and Sun tinctures, the Lunar and Solar forces that they have brought into earthly incarnation. We note that the Lunar stands nearer the male partner and the Solar is seen close to the female alchemist.

Below this again is seen a dying Hercules figure, complete with his lion's skin and club. Although it might initially seem to be a digression, I feel it is important to remind ourselves of some aspects of the Hercules myth, as it is an important underlying structure to this alchemical series. Hercules was the son of the god Zeus-Jupiter and the mortal Alcmene. Hera-Juno, the wife of Zeus, was of course jealous of this relationship and placed Hercules under an obligation to perform a series of twelve labors for King Eurystheus. These prodigious labors required Hercules as hero to work entirely through the outward, aggressive male facet of his being; however, once these were completed, Hercules underwent a different experience. For one of his outbursts of temper, he was made to serve Queen Omphale for three years. While in this service, the hero's nature changed. He lived effeminately, wearing at times the dress of a woman, and spinning wool with the hand-maidens of Queen Omphale, while the Queen took to wearing his lion's skin. Hercules here experienced his inner, feminine side. After this, he married Deianiera and lived at peace with her for three years. Later, an important incident occurred when he killed the Centaur Nessus. The dying Centaur told the gullible Deianiera that a potion made of his blood could be used as a charm to preserve the love of her husband for her. This was the undoing of Hercules, as this blood was, in fact, the most potent poison. Deianiera steeped a white robe in this blood and gave it to Hercules. The poison of the blood entered through his skin, causing him the most intense pain. He tore off the garment and pieces of his skin came away with it. Knowing his end was near, he prepared to die and ascended Mount Oeta to build himself a funeral pyre. There he laid himself down with his head resting upon his club and his lion's skin spread around him. As the flames of the funeral pyre engulfed him and consumed his mortal part, Zeus claimed his immortal part, and welcomed him to the abode of the Gods. Zeus enveloped him in a cloud and took him up in a four-horse chariot to dwell among the stars. Hera-Juno became reconciled to Hercules and gave him her daughter Hebe in marriage.

This myth parallels some facets of the alchemical process outlined in the *Mutus Liber*. We see the mortal part of Hercules dying at the bottom of our final illustration, and the immortal part (the pair of alchemists) being raised up to heaven by Zeus. Hercules' death is preceded by a period during which he inwardly encounters his feminine side and thus balances his life's energies, which until that time had been polarized in the male, active facet of his being. There is a deal more to be learned of the alchemical process by contemplating this myth in depth, though we shall not pursue this any further.

In view of the Hercules connection, and his Eleventh Labor of the Golden Apples from the Garden of the Hesperides, it may not be too fanciful to identify the boughs forming the oval in our last illustration as the apple tree. The apple is also a member of the rose family and thus the apple fruit is a perfect completion of the rose blossom of the first illustration.

At the base of the illustration where the two branches meet and are tied together, two feathered wings are seen indicating the spiritual ascent. Below, at the very base, is another small piscina or oval wreath of leaves, within which is seen a shield. I suggest the following tentative interpretation of this alchemical coat of arms: Perhaps this also is a picture of the alchemical process. At the bottom we have the earthly substance

which rises in three waves. This is the inner potential form embodied in the lowest matter. In the next level these three nodes of potential have been purified and separated out into three spheres or globes, and they lie below the chevron which separates the higher spiritual realm from this intermediate realm of the psyche. Above the gulf of the chevron, we see three shells, indicating the highest spiritual manifestation of the process. Perhaps the shell is an excellent symbol for the Spirit working in dense matter, as the shell is formed by the laying down of substance within an archetypal form. Shells grow like plants, but are composed of hard, mineralized substance. These three levels correspond to the three levels of Spirit, Soul and Material in the larger illustration above.

So our final plate, marking the end of the Art of this alchemical process, is the attainment of a spiritualization. The subtle part of the soul of the alchemist is able to ascend to communicate with the spiritual world, just as the immortal part of Hercules was able to enter the realm of the Gods, the starry heavens.

The Three Initial Plates of the Process

Now let us consider the three initial illustrations of the three stages we propose in our scheme of analysis, these being plates 2, 8 and 11 (shown together on the overleaf). They are very similar in form, though there is a definite development and progression in their symbols.

Each figure is divided into an upper and a lower section. In the lower section, we see the outer, earthly manifestation of the process in the alchemists' chamber, while in the section above, the spiritual forces at work in the alchemists' vessel are exhibited. Thus these two sections are united in the common symbol of the sealed flask.

In the sections below, we see at the center a furnace in the stylized form of a castellated tower. There are three levels to this furnace. At the base, in an arch between two small columns, a lamp

COMMENTARY

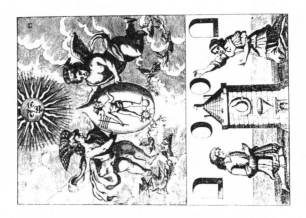

Plates 2, 8 and 11.

is burning to provide heat for the process. Just above this, in the center of the furnace, a funnel-shaped structure is seen; and at the top level of the furnace, in an egg-shaped enclosure, a sealed flask is seated upon a ring. It seems likely that what is intended, on the physical alchemical level, is that the flask is to be gently heated upon a steam (or perhaps sand) bath, the funnel structure containing the water (or sand) to spread the heat. We must note that in plate 8, the initiatory figure of the second process, no lamp has been set, and consequently this stage must develop with its own internal heat. On the left side of the furnace, the male alchemist is pictured, and on the right, his female partner. The male is, in all three figures, shown in an attitude of prayer, while his female counterpart is more active, gesticulating and perhaps even making some statement. This indicates a passive meditation, a listening to the inward soul, for the male alchemist, and an externalization of inner psychic contents, as prophecy or inspirational sayings, on the part of the female alchemist. (Interestingly, this was precisely the relationship that Armand Barbault and his wife adopted when working alchemically with the *Mutus Liber*). One interpretation of this would be that these two figures, the male and female partners, are in fact parts of the individual soul of the alchemist undertaking this work. However, equally valid is the viewpoint that the Great Work cannot be achieved except by such a partnership of the alchemist and his Soror Mystica. The idea of the Mystical Marriage and the Chemical Wedding is an important element in the esoteric alchemical tradition. In a sense, both of these viewpoints are simultaneously valid, and I will try to incorporate both of them into this commentary.

At the first stage, plate 2, heavy drapes almost obscure the background behind the two figures. However, in the next stage, plate 8, these are somewhat drawn aside to reveal two windows; and finally, in plate 11, these drapes are entirely removed, and two further oval windows allow the pair of alchemists to view outside their chamber.

What they are able to view, as their spiritual vision evolves, is that which is shown in the upper part of the figures, for here the spiritual forces at work in their flasks are revealed for us to see.

At the beginning of the first process, the flask is seen to contain Neptune with his trident seated upon a dolphin throne, his arms reaching out to embrace the two smaller figures of the Sun and Moon. At this stage of the process, the work must be done in a watery medium, which Neptune of course controls. Thus we note that the Angel supporters of the flask (these Angels indicating the spiritual forces involved in the work) stand with their feet in a watery sea below. However, as we move to the next stage we are entirely in an airy realm, having ascended far above this watery sea. This is shown by the birds below the feet of the Cherubim who support the flask. Within this flask, Neptune is no longer seen, but now Mercurius appears with his winged cap and caduceus rod. He stands, at the start of the second stage (plate 8), upon an earthy substance where the Sun and Moon archetypes are visible. Thus the watery realm within which the Sun and Moon were dissolved has now dried up, giving rise to two layers—the earthy mass with undigested Sun and Moon, and the airy realm in the flask where Mercurius makes his home. Ten birds are seen, in two groups of five. The leader of the group on the left, immediately above the male alchemist (and on the solar side of the flask) bears the gift of 𓎛, Philosophical Sulphur to the process; while on the right, above the woman alchemist (and the lunar side of the flask), the bird leader bears a branch with ⚹ , Sal Ammoniac. These two substances, which are also archetypal energies, are important for the development of this process. In plate 11, when the process has undergone a further stage in its evolution, the earthy mass at the base of the flask is no longer to be seen, and Mercurius stands in the flask directly upon the Sun and Moon archetypes. A stage of purification from the earthy dross has occurred, and the archetypes freed into their more subtle airy form. The birds this time bear the gifts of 𓎛 , sublimated Philosophical Sulphur, and again ⚹ , Sal

Ammoniac. Sal Ammoniac (ammonium chloride) is important in
alchemy because it sublimes; that is, on being heated in a flask it
moves from the solid to a gas without going through an intermediate liquid stage, and then condenses on the cooler parts of the
flask, moving from gas to solid, again without an intermediate
liquid state. Thus it symbolizes or typifies for the alchemist any
subtle or 'sublime' substance which seems to bear within it great
spiritual potential, moving easily from spirit to matter. To the
alchemists, the term Sal Ammoniac was not merely ammonium
chloride, but was more broadly applied to a potential state of all
matter, and an archetypal force at work in all matter.

Thus through these three illustrations we see an evolution, a
subtilization of substance through the work of the individual
stages which they initiate. We shall now look at each of these
stages in some detail.

The First Process (Plates 2-7)

This begins with plate 2, already described in some detail, and as we have seen, this first process will be initiated in the watery element under the power of Neptune. Our alchemical couple must spiritually find the solar and lunar forces and dissolve them in this watery realm.

Plate 3 indicates a most important facet of the work. Initially, one might feel this plate to be out of sequence, having connections through its use of classical mythological symbols with the first and last plates of this series, but it seems to be placed here to point to the source of the prima materia the alchemists must seek for this process. Here is depicted the globe of the Earth divided into three regions—the realm of the air above the Earth, the surface of the Earth, and the watery realm below the surface. Outside of this globe is seen the Cosmic-Spiritual world with its Solar and Lunar

Plate 3.

forces, and above the globe Zeus-Jupiter is depicted upon his eagle watching over the lower world. The surface of the Earth we inhabit is thus seen as the interface of two realms: there are forces working from above downward, pictured as Juno and her peacock and the birds in the air; and forces working from below upward, pictured as the realm of Poseidon-Neptune with the fish and mermaid or melusina creature. Living on the surface of the Earth, man lives between these two streams of force.

These are the currents of the Ethers that work in the Seasons: the Upper Ethers, the Ouranian forces, which descend down to the surface in the Spring and Summer and are held back in Winter; and the Lower Ethers, the Chthonic forces, which rise up to the surface from the depths of the Earth in Spring and Summer, and descend again in the Winter. We also note that this picture corresponds to the soul of man. For man's soul, his psyche, lives between consciousness (paralleling the light of the Ouranian forces) and unconsciousness (the dark Chthonic realm). The psyche is that delicate soul membrane between these two realms which it needs for its proper nourishment to partake of both realms. Thus we have the phenomenon of the cycle of sleep and waking consciousness (like the seasons on the Earth), and of rational understanding and dreamy inspiration.

We can approach this figure of the *Mutus Liber* on either of these levels, as an indication of the forces at work in the Earth or in the inner, soul-realm of man. On the Earth's surface, we see most importantly a ram and a bull at opposite sides of our globe (ram-solar and bull-lunar). These two symbols will appear in later plates. Above, a female Earth Mother figure tends her delicate plants, one group planted in the earth and the other artificially cultivated in a pot, thus showing that the task of alchemy is to artificially replicate a natural process occurring in the earth. An attempt is being made by the couple below this to capture the subtle, living energies of the domains above and below the earth's surface, the task of alchemy being, as we have noted, to work with the Ouranian and Chthonic Ethers (and the conscious and unconscious forces in the soul).

Now, moving on to plate 4, the third in this outline of the process, we see the picture most often reproduced from the *Mutus Liber* series. Our alchemical couple is gathering dew, which has been deposited on cloths they have suspended on pegs horizontally just above the ground. This is taking place on a meadow where the ram and the bull are seen, which here may signify the adjacent as-

Plate 4.

trological signs of Aries and Taurus; and because these are spring signs (together spanning the period from the spring equinox March 21st-May 20th), they probably indicate that the dew must be collected in spring. Aries is Cardinal Fire and Taurus a Fixed Earth sign, and therefore we have a meeting of two opposites. The ram may also indicate the stream of Ouranian forces, and the bull the Chthonic earthly energies. Streams of force descend from the Cosmos overhead, while Sun and Moon shine above their respective charges. The important element of this figure is the dew that is being collected by our alchemical couple. Dew is a very special substance. It does not arise merely from a precipitation of the atmosphere like rain or the dampness of a mist, but by a subtle distillation process. For the formation of dew, a warm earth is required, a clear sky at night and cold surface air. Water in the form of dew is distilled from this warm earth, from the soil and vegetation, and condenses on the surfaces of leaves or other cold objects just above ground level. This dew contains various essences taken from the earth and the vegetation during its formation. It also bears within it a richness of etheric force.

The biblical quotations on plate 1 initially draw our attention to the importance of the dew in this work.

> May God give you of the dew of heaven, and of the fatness of the earth, and plenty of grain and wine. (Genesis 27:28)
>
> Then Isaac his father answered him: "Behold, away from the fatness of the earth shall your dwelling be, and away from the dew of heaven on high. . . ." (Genesis 27:39)
>
> And of Zebulun he said, "Rejoice, Zebulun, in your going out; and Issachar, in your tents. . . ." (Deuteronomy 33:18)
>
> So Israel dwelt in safety, the fountain of Jacob alone, in a land of grain and wine; yea, his heavens drop down dew. (Deuteronomy 33:28)

Our alchemical couple take this dew, which they have wrung

out into a shallow pan, and work with it in their laboratory through a series of operations. This is depicted in plates 5, 6, and 7, like a strip cartoon, and I will now examine it in sequence.

We begin with our couple placing their dew into a vessel set into a furnace, and in the next frame the female partner places a still head upon this vessel, while the man holds a round receiving flask to the outlet. The furnace having been lit, a volume of distillate is shown collecting in this flask. In the next picture, the woman spoons the residues, which are here given the symbol ⌂⌂⌂, into a square-bottomed bottle, which in the next frame she hands to a Kronos-Saturn figure, who is depicted in characteristic pose about to devour his child. Beside him on a shield, the symbol Luna ☽ is shown; this also appears on his chest, and he stands upon a slight mound of earth. We will see later how this lunar aspect hidden within him is eventually purified and released.

Thus through these four stages our dew has been, through a distillation, separated into two fractions, and we could write the following equation:

$$\text{Dew} \left\langle \begin{array}{l} \text{First Distillate in round-bottomed flask} \\ \\ \bigtriangleup\bigtriangleup\bigtriangleup \longrightarrow \text{Saturn} \end{array} \right.$$

The next stage shows the First Distillate being poured into four cylindrical vessels fitted as cucurbites; that is, for continual reflux, the condensed vapors run back down into the liquid. This is the

alchemical process of Circulation, and here it is undertaken in four separate parts, simultaneously heated by a furnace, which provides an indirect and temperate heat. This circulation should be continued for 40 days, and during this time an inner change should occur in the distillate.

The next sequence of illustrations shows the removal of this transformed distillate from the four circulating vessels and it being

poured into a flask, which in the first frame above is placed upon a water bath in the furnace. The second frame shows this being heated on the water bath, which provides a gentle and constant heating about the boiling point of water. The female alchemist has fitted a still head, and a further distillation occurs, the distillate being collected in a round-bottomed receiver held by her partner. During this heating, a six-petalled flower forms in the distilling flask, and once this has fully formed, the furnace is cooled and the woman is seen removing this from the reaction vessel and placing it into a round bottle, which in the fourth frame is handed by the man to Apollo. We can write an equation for this process as follows:

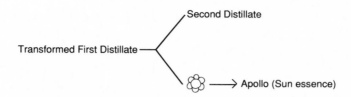

The next stage begins with the female alchemist taking the Saturn residues △△△ and heating them to dryness in a shallow crucible set upon the furnace. The second frame then shows her

pounding and grinding these dry residues in a shallow dish, while the male alchemist pours upon this the contents of a round-bottomed flask. This must be the Second Distillate. This mixture is then poured into a large round-bottomed flask and presumably left for some while to digest.

Later, in the fourth frame of this stage, this is poured into a shallow vessel set upon a water bath and gently evaporated, the

residues, labeled ✳, being collected by the woman alchemist and placed in a square bottle. We could notate this as follows:

The following strip shows this process allegorically. In the first image we see Saturn-Kronos (the ⌂⌂⌂ residues) being heated

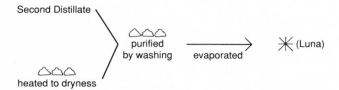

upon a fire. The man then pours the contents of the round-bottomed flask upon Saturn in his tub, and arising from this he is able to free himself of the ✳ material—thus we see him in the act of cutting the cord binding him to this with his sword. The ✳ is presented to Luna.

So to examine the whole process in diagrammatic form:

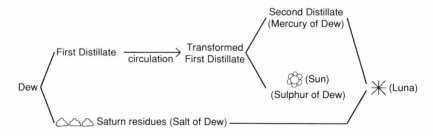

Thus we have a classical spagyric technique, separating and joining together the essences of the primal material, the Dew. One could describe the ⌂⌂⌂ residues as the SALT of the Dew, the first

Distillate being a crude mixture of the Mercury and Sulphur of the Dew. Through the digestion process (Circulation), the Sulphur of the Dew ✿ separates out from the Mercury which is distilled off as the second Distillate. The Mercury is then united with the Salt and a tincture ✳ formed. At the end of this experiment, we have only ✿ (Solar) and ✳ (Lunar) left.

Now, if we treat this process entirely physically, at face value, it will be obvious that large quantities of dew will have to be collected in order to obtain sufficient quantities of the residues; and to the best of my knowledge, no one has repeated this experiment exactly as outlined. Armand Barbault showed that such large quantities of dew could be collected, and with considerable labor and dedication to the task, was able to obtain some hundreds of liters in a season. As I have indicated, dew, through its formation, contains various dissolved substances and essences of the plant world and the soil it is breathed out from. We might note here a kind of parallel with the preparation of the Bach Flower Remedies, which are made by placing flower heads upon the surface of a vessel of pure water in bright sunlight. Extremely small quantities of the subtle essences of the plant are transferred to the water, which is then used as the Mother Liquor for the Flower Remedy.

One possible interpretation of the *Mutus Liber* is that 'dew' is not necessarily restricted to one physical substance, but anything that is seen to have been formed in a similar way may be taken as the prima materia. Thus the important element in this is that it would be breathed out and distilled from earthly substance. Thus dew, as the nectar or essence of the earth, could perhaps be transferred to any plant essence, the fragrance of flowers, or the essential oils and resins which are breathed out of plant substance. It could also be applied to some minerals which seem to form dews on their surfaces—the deliquescent substances, such as Tartar. This is merely another possible interpretation of the work, which may be of value to some readers.

If we examine this process etherically the following picture emerges. The dew contains a rich etheric energy, having been formed by the uprising etheric forces in the Earth. Here we may note the importance of site. Through my own researches into the etheric energies in the Earth (see my article "The Alchemy of the Earth Forces" in *The Hermetic Journal*, No. 10,), I have confirmed that there are certain spots on the Earth particularly rich in these upwelling etheric forces. Such areas were often recognized as sacred by the ancient peoples and are in consequence important archaeological sites, often marked by standing stones, stone circles, cairns or henge monuments. The activity of such sites can be detected by dowsing, and it would be best if one wished to work with the etheric implications of the *Mutus Liber* to collect one's dew at such an etherically active site. Armand Barbault located one such area rich in etheric force by following the intuitive perceptions of his wife and partner in his alchemical work.

According to some esoteric traditions there are two Ethers wrapped in this stream of upwelling force—the LIFE ETHER and the TRANSFORMING ETHER. The Life Ether corresponds to the Earth Element, and the Transforming Ether is the etheric counterpart of Water. (The Ouranian Ethers that play in from above are the LIGHT ETHER, corresponding to Air, and the WARMTH ETHER, paralleling the Fire element.)

During the first distillation, the Ethers woven into the dew pass into the First Distillate, together with the more subtle physical substances in the watery vehicle; left behind in the vessel is a purely physical residue stripped of its etheric force. The Circulation stage digests a part of the etheric force, the Life Ether that is committed more to the Earth element, and this works its pattern into the small quantity of subtle physical substances dissolved in the First Distillate. During the next stage, when heated upon the water bath, this Life Ether becomes committed to incarnate in these substances, and this is seen in the formation of the flower. This flower is woven by the Life Ether forces, using these subtle substances in the watery menstruum as a vehicle. The Transforming Ether distils off into the Second Distillate. Later, this is reunited with the physical residue, a △△△, and this Transforming Ether (also called in some traditions, the Chemical Ether) changes the dead △△△ into living ✳. Thus we have, in ⊛ and ✳, the physical vehicles for the incarnated Life Ether and Transforming Ether found in the original dew.

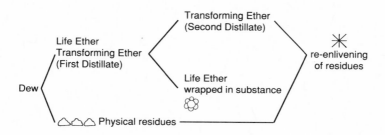

This picture of the etheric dimensions of the process will allow us to grasp more readily some facets of the later stages of the *Mutus Liber* process.

The Second Process (Plates 8-10)

The next phase of the work begins with plate 8, which we have examined earlier. This phase sets the spiritual direction of the process, and we now understand a little more as to why the birds bear gifts of ⚶ and ✳ to the flask. Now, ⚶ is the alchemical symbol for Philosophical Sulphur, which corresponds as we have seen to the ⊛ Solar essence of the first process, while ✳ is, of course, our previously gained Lunar essence. These must now be brought together in the flask of transformation.

During this stage, our alchemists below keep their furnace without a lamp, thereby allowing the process to occur at its own inner heat. However, having achieved the First Process, a little more light creeps into their chamber, as the drapes have been drawn back slightly, and a little more illumination enters their souls.

Plate 9, although echoing plate 4 in form, shows a radical departure. Instead of six cloths spread out on frames upon the meadow to catch the rising dew, six large plates are placed in a similar arrangement. These shallow vessels would not be of much use in gathering the rising dew, but would collect any precipitation descending from above, and perhaps in this instance rain water or condensed water vapor from the air is intended to be collected. Again, this is undertaken when the ram and the bull consort on the meadow. The female alchemist pours the collected precipitation into a large round-bottomed flask which she hands to Mercurius.

The Second Process begins in the first frame of plate 10, where the alchemical couple take equal parts (measured in the pans of the

balance) of their essences ✳ and ✿ , and place these into a round-bottomed flask. The second frame shows the man pouring into this the contents of a round-bottomed flask, and this can only be the precipitation collected in the previous plate, plate 9. This flask is hermetically sealed by the male of the alchemical partnership, and together they place the flask into the furnace of transformation,

where it is to be slowly heated. Thus a conjunction of Sun and Moon is brought about, as is shown below. The furnace has beside it a cross-section of an egg, which has four layers to it.

To follow up our etheric picture, the precipitate descending from above brings with it, woven into its substance, the pattern of Ouranian Ethers—the Light Ether and the Warmth Ether. (This inherent pattern can be seen in the formation of snowflakes, which have a regular six-fold form but a vast multiplicity of individual crystalline structures—the archetypal pattern being a result of the working of the Ethers, while the precise manner that this is articulated in substance is a result of the material forces at work in the water vapor.) These Ouranian Ethers are in an unfixed form in their watery vehicle, and are united with the embodied Transforming and Life Ethers prepared in the earlier process. Through this digestion an etherically rich menstruum is formed, joining together in a harmonized way the four ethers. This would undoubtedly be an effective remedy for disharmony in the etheric body of man, and should also help in treating conditions resulting from disturbances emanating from an imbalance of soul energies.

Until this point things seem quite clear, but now the *Mutus Liber* becomes obscure and hides its intentions. As I pointed out in the introduction, the earlier (1677) version of this work includes only 13 plates. Two plates have thus been added to the Manget edition (or left out of the Saulat), these being Manget's plates 12 and 13. We note that they are almost identical with plates 9 and 10, and we are, in consequence, posed the difficult question of whether to include them or not in our interpretation.

I intend to avoid this question by providing viewpoints which include both of these cases.

Let us first assume that plates 12 and 13 are part of this sequence. Then we see that the sequence 11-12-13 is a kind of recapitulation of 8-9-10. In alchemy there are no such mechanical repetitions of processes, and we have here a kind of higher order conjunction of the elements of the work. Plate 11, which we have looked at before, has some important differences from plate 8. The flask borne aloft by the cherubim contains Mercurius, who now stands free within this vessel from the earthy dregs, which we see at the bottom in plate 8. The Sun and Moon are entirely dissolved or digested into, this Mercurial menstruum. We note further that the birds bear 添, *Sublimated* Philosophical Sulphur, as well as ✳ to the flask.

When we move to the first frame of plate 12, we see, as in plate 9, the alchemists weighing out equal quantities of substances to

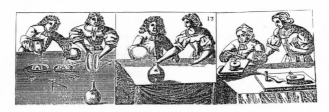

place in their vessel. Here these are ✳ and ✴. Now we know ✳ from before, but what is this ✴ ? The only thing internal to this operation it can be—that is, assuming no outside substance is at this stage brought into the process—is the outcome of the first conjunction, namely that which was created in the furnace of transformation in plate 10: the conjunction of ✳ and ⊛ in the Mercurial water. This is now wedded to an equal quantity of ✳ , and mixed with a further measure of the precipitation gathered in plate 11. After this is sealed and placed in the furnace, the outcome of this process is shown in the conjunction in the last frame of plate 13, where we note the figures 100 . 1000 . 10000 . etc. In plate 10, only the figure 10 appears. Now, we can tentatively suggest that here we have a process paralleling that of trituration in the preparation of homoeopathic medicines. That is, the first Conjunction requiring equal parts of ✳ and ⊛ is mixed with 10 parts of the Mercurial water to form the first dilution (denoted as D1 in homoeopathy). At the next stage, equal parts of ✳ and ✴ (the first order dilution) are digested together with 10 parts of the Mercurius, thus forming the second dilution (denoted 100 or D2), and further cycles of this process produce in turn a dilution to 1000th (D3), 10,000th (D4), etc. This is the well-known process of Multiplication in alchemy, which is paralleled in homoeopathic trituration, the rhythmic dilution of substances into infinitesimals. In homoeopathy, as the material body of the substance is diluted, its etheric energy is released. Therefore the higher dilutions, which have no physical molecules of the original substance of the remedy in them, are etherically very powerful and are used in treating acute cases to produce a sudden change in the patient's condition. We note that in homoeopathy this dilution or attenuation is usually made in the solid phase, using an inert substance like sugar to perform the rhythmic decimal dilutions, and the word 'trituration' indicates this process of grinding. In the *Mutus Liber* process, the work uses the liquid phase, the Mercury in the form of the collected rainwater or other precipitation from above.

Let us look at the etheric parameters of this process of further conjunctions. We have at the beginning of our first dilution ⚗ the Sulphur of the Dew (which is solar in quality) and ✳ the Salt of the Dew (lunar in quality). The Sulphur was formed subtly in this process, like a delicate flower in a flask of already subtle distilled material. The Salt ✳ was formed by an enlivening of the Saturnine lees △△△ left after the first distillation (plate 5). Thus we will have, in quantity, more of ✳ than of ⚗. (Note that ✳ is in a square bottle, while ⚗ is placed in a round-bottomed vessel.) The ✳ (the symbol suggests a union of ✳ and ⚗) which we have as the result of this first conjunction, therefore, has qualitatively more Sulphur energy than Salt, and requires further additions of ✳ , and subsequent digestions in the menstruum of Mercurius within the furnace of transformation to form it into an entirely balanced remedy. We could, therefore, also see 11-12-13 as a series of operations repeated cyclically until all the ✳ has been absorbed. As this requires a period of digestion, each of these conjunctions would have to await a new year, a new spring.

If we, alternatively, assume that plates 12 and 13 of Manget's *Mutus Liber* are redundant, then the tincture formed at that point of the process described in plate 10 is the final preparation and contains, in itself, all that is necessary for the further development of the process. We would now move directly from plate 11 to the penultimate figure in the series, plate 14 in Manget's edition.

Here we see the same three furnaces as are shown in the alchemists' chamber pictured on plates 2, 8 and 11. Below them we note three figures trimming their lamps and filling them with oil,

adjusting these for the process which will take place in the three furnaces.

These figures are Man, Child and Woman, and beside them in Roman numerals are VI, II and X, which here probably indicate the number of months that each of their processes should be continued. The first process, under the rule of the Man-Father, is surely the separation of the Dew into its essential principles; the next process, under the rule of the Woman-Mother figure, must be the conjunction and integration of the solar and lunar principles (Sulphur and Salt of the Dew) in the Mercury; while the process corresponding to the Child is only hinted at in the third frame below.

The final task of our alchemical couple must be to form a 'Stone' from the liquid Tincture left after the previous stages. This is placed into stout vessels and heated upon a strong fire. One can only surmise here, taking our inspiration from Armand Barbault, that the final task of the alchemists was to transfer the energies of the watery tincture to the solid phase, and for that this finely powdered Silver and Gold acted as mediums for fixing the tincture. We see, between the furnaces for this work, a balance, and below, a mortar and pestle, in which the metals can be reduced to powder.

Our alchemical couple appear in the last frame of plate 14, to tell us to 'Pray, Read, Read, Read, Read again, Work and you shall

find.' The task of alchemy begins with prayer as it is a spiritual work, then one must read one's texts and material three times; that is, once as a physical task, next as a Soul Alchemy, and then as Spiritual Alchemy, then read again uniting these three dimensions. Only then can one begin the work and find the goal of alchemical transformation.

Being aware of this threefold nature of Alchemy, we should be cautioned against accepting the interpretation outlined here as the only possible one, and recognize that there are other ways of approaching the work. We will now summarize some other possible perspectives on the *Mutus Liber*, other questions posed us by this mysterious book of symbols.

We can ask whether the male and female alchemists are indeed intended as separate individuals or are merely parts of the one person. We could look at all the different actions of the two partners as occurring in separate parts of the one personality, and could, for example, attach symbolic meaning to these actions.

Thus the male often holds the receiving flask for distillate, and the woman controls the still head in the sequence of operations, and we could pursue an interpretation which ascribed a masculine or feminine dimension to such operations.

On the other hand, it may well be that this process outlined in the *Mutus Liber* indeed requires such a partnership of a man and a woman for its success. We might be tempted to read into this document various parallels with the Tantric practices of the East (and we should note that these are not merely the often sensationally-reported sexual rituals, but that Tantra is a much broader realm of spiritual science, akin to alchemy in the West). However, while I feel that a Tantric interpretation may be relevant here, the *Mutus Liber* does not present its symbols in a way which is straightforward to interpret Tantrically. However, the various pieces of apparatus—retorts, flasks, furnaces—can be seen as metaphors for a union of the male and female. Also with the meadow upon which the ram and bull are sporting we have a concern with fecundity, and because, as I have indicated in my interpretation of the dew sequence, there is a connection with the ethers and the cycle of the seasons, it may not be too far-fetched to draw such a parallel with Tantricism. However, I doubt whether the author of the *Mutus Liber* intended us to work entirely on this level, and I have not pursued this level of interpretation to any length in this commentary.

The question also arises as to whether the gathering of the dew is not some kind of metaphor for an inner process. In such a Soul Alchemical view of the *Mutus Liber*, the dew would be seen as that which rises into the psyche out of the dark, unconscious sphere within us. This material must be taken hold of and inwardly transformed, or else the soul stands in some peril of being swamped by such unconscious psychic material. The alchemical couple working with the dew are *animus* and *anima* coming to terms with the impulses arising from the unconscious, and the *Mutus Liber* symbolizes a way of working with this through a Soul Alchemy.

Johannes Fabricius, in his *Alchemy: The Medieval Alchemists and their Royal Art* (Copenhagen, 1976) introduces a strongly Freudian interpretation, in which the couple (here brother and sister) are not gathering dew so much as washing the filthy sheets of their incestuous love; he sees the process as a cleansing from the various psychological difficulties (Oedipus complex and all) seen in child-hood, adolescence and maturity, through the eyes of a Freudian psychologist. I feel Fabricius' interpretation to be rather limited, and that it barely touches upon the great mysteries of the *Mutus Liber*; however, he does present a coherent view of the work, which might be of interest to some students.

We must resist the temptation to reduce this work to a one-dimensional interpretation, but instead glory in its multiplicity of facets. Thus, for me, the *Mutus Liber* simultaneously outlines the simple spagyric technique for preparing the Salt, Sulphur and Mercury of substances; a working with the Ethers in the cycle of the seasons, and incarnating their essences in substances derived from dew; the preparation of an Etheric Medicine; a process of inner development through a Soul Alchemy, involving the union of masculine and feminine (solar and lunar) facets of our inner being; a Spiritual Alchemical task which involves working with a part-ner; and a process for preparing homoeopathic tinctures, medicines for the soul and body. All of these, and others too, are facets of this work, and they all, in a sense, resonate together. To narrow the work down to any single interpretation would be to destroy this harmonic resonance. For it is in this multiplicity that the inner potential energy of the *Mutus Liber* resides. Working with its symbols can unlock many different facets of alchemy for us, which we should try to hold before us simultaneously, realizing that this archetypal process works on many levels at once.

Thus, the *Mutus Liber* reveals itself as a valid document of alchemy still relevant to us today, which is well-pictured as being under the guardianship of Hermes-Mercury, the most quixotic and manifold among the gods.

Magnum Opus Hermetic Sourceworks

The Magical Calendar
The Mosaical Philosophy - Cabala
The Crowning of Nature
The Rosicrucian Emblems of Cramer
The Hermetic Garden of Daniel Stolcius
The Rosary of the Philosophers
The Amphitheatre Engravings of Heinrich Khunrath
Splendor Solis
The 'Key' of Jacob Boehme
The Revelation of Revelations of Jane Leade
A Commentary on the Mutus Liber
The Steganographia of Trithemius
The Origin and Structure of the Cosmos
A Commentary on Goethe's Fairy Tale
A Treatise on Angel Magic
The Paradoxical Emblems of Freher
The Heptarchia Mystica of John Dee
The Chemical Wedding of Christian Rosenkreutz
Alchemical Engravings of Mylius
The Five Books of Mystical Exercises of John Dee
Atalanta Fugiens of Michael Maier
The Kabbalistic Diagrams of Rosenroth

In addition to issuing the Magnum Opus Hermetic Sourceworks series, PHANES PRESS both publishes and distributes many fine books which relate to the philosophical, religious and spiritual traditions of the Western world. To obtain a copy of our current catalogue, please write:

PHANES PRESS
PO BOX 6114
GRAND RAPIDS, MI 49516
USA